A Question of Math Book

Addition

by Sheila Cato
illustrations by Sami Sweeten

Carolrhoda Books, Inc./Minneapolis

This edition published in 1999 by Carolrhoda Books, Inc.

Carolrhoda Books, Inc., c/o The Lerner Publishing Group
241 First Avenue North, Minneapolis, MN 55401 U.S.A.

Website address: www. lernerbooks.com

LIBRARY OF CONGRESS CATALOGING-IN-PUBLICATION DATA
Cato, Sheila.
    Addition / by Sheila Cato : illustrations by Sami Sweeten.
        p.   cm. — (A question of math book)
    Summary: A group of children introduce addition, using everyday examples and practice problems.
    ISBN 1-57505-320-9 (alk. paper)
    1. Addition—Juvenile literature. [1. Addition.]   I. Sweeten,
Sami, ill. II. Title. III. Series: Cato, Sheila, 1936-   Question
of math book.
QA115.C25 1999
513.2'
11—dc21   98-15725

The series A Question of Math is produced by Carolrhoda Books, Inc., in cooperation with Brown Packaging Partworks Limited, London, England.
The series is based on a concept by Sidney Rosen, Ph.D.
Series consultant: Kimi Hosoume, University of California at Berkeley
Editor: Anne O'Daly
Designers: Janelle Barker and Duncan Brown

Printed in Singapore
Bound in the United States of America

1 2 3 4 5 6 - JR - 04 03 02 01 00 99

Here's Josh with his number friend Digit. Josh is learning about math the fun way. Digit is going to help Josh find the answers to some addition questions, and you can join in too. You will need colored blocks, beads, buttons, pennies, a deck of cards, some paper, and colored pencils.

I've learned the numbers from 1 to 10. Is there a fun way to practice putting them in the right order?

Of course, Josh. It's always good to practice using what you've learned. Why not start by hanging these T-shirts on the clothesline. Each T-shirt has a number on it.

4

Special Signs

Traffic lights are signs that tell drivers when to stop and when to go. There are special signs in math, too. The sign + means add or plus. The sign = means equals.

Find the T-shirt with the digit 1 on it. Pin that on the line first. Now look for the T-shirt with the digit 2. I'll leave you to do the rest.

That's easy! We'll hang all the T-shirts on the line and we can practice saying the numbers in order to make sure we remember them.

1, 2, 3, 4, 5, 6, 7, 8, 9, 10

Now You Try

Take the face cards out of a pack of playing cards and set them aside. Give the red number cards to a friend and keep the black number ones. Have a race to put the cards in the right order.

Oh, no! My 6 pet mice have escaped from their cage and they're all over the room. I need to get them back into their cage. How can I be sure that I've caught them all?

That's quite a problem. It's hard to count mice when they're running around. You could count each one as you put it back in the cage. Then you'll know when you've found all the mice.

That's a good idea. There's a mouse on the blinds, so that's 1. There's another mouse running up the table leg, so 1 and 1 is 2. There's one on the top of the cage, 2 plus 1 is 3.

One by One

When you add 1 to a number, the answer is always the next number. 2 + 1 = 3, and 3 is the number that comes after 2. 3 + 1 = 4, and 4 is the number that comes after 3.

I'll keep on adding the mice, one by one, until I have found them all.

You could write special sentences called equations, using the math signs, to show how many mice you have caught:

$$1 + 1 = 2$$
$$2 + 1 = 3$$
$$3 + 1 = 4$$

Thanks, Digit. In the future, I'll remember to keep the door of the cage locked!

Now You Try

Cut out pictures of 10 people from a magazine. Paste them in a line on a big piece of paper. Underneath each picture, write the numbers from 1 to 10.

Holly and I have found some colored beads in my toy box. There are 2 red beads, 2 yellow beads, and 2 blue beads. What can we make with the beads?

Why not make a necklace? Find a piece of string and thread on 2 blue beads, then 2 red ones. Now there are 4 beads on the string. Thread on 2 yellow beads and count the beads again. How many beads are on the string?

There are 6 beads. I'll thread on 2 green beads. That makes 8. To finish the necklace, I need 2 orange beads. Now the necklace has 10 beads.

Adding the beads on 2 at a time makes the sums

$$2 + 2 = 4, \quad 4 + 2 = 6,$$
$$6 + 2 = 8, \quad 8 + 2 = 10$$

If I push the beads together and tie the ends of the string, the necklace is ready to wear.

Sets or Groups

A set is a group of things. The things can be animals, people, or beads, or anything else. When you join 2 sets together, this is called addition. The answer is called the sum.

Now You Try

If Josh and Holly had added 2 more beads, how many beads would there be on the necklace?

The apples on the trees in my garden are ripe. Brad is helping me pick them. Brad has picked 4 apples and I have picked 3 apples. How many apples have we picked all together?

12 beads

Hold out the apples so that you can see them. Count the apples you are holding, then keep on counting the apples Brad has picked.

I am holding 1, 2, 3 apples and Brad has 4, 5, 6, 7. We have picked 7 apples all together. I can write this as an equation

$$3 + 4 = 7$$

I'm going to ask my mom to make an apple pie!

Practice Makes Perfect

The best way to learn addition is to practice adding! Look for things to add in your room, in your kitchen, at school, and in the garden. Everywhere you go, there are things to add!

Now You Try

If Brad had 5 apples and Josh had 3 apples, how many apples would the friends have all together?

11

Here I am in the kitchen. Mia is helping me bake some cookies to share with our friends. I've made 4 cookies and Mia has made 3 cookies. What's the quickest way to find the total number of cookies?

You know that you have made 4 cookies, and you know the numbers up to 10. Keep on counting 3 more from 4 and that will give you the total number of cookies.

The number after 4 is 5, the next number is 6, and the number after that is 7. Mia and I have made 7 cookies.

$$4 + 3 = 7$$

There's plenty of cookies to share with our friends, and Mia's dog can have one too!

Which Order?

Josh and Brad picked 3 + 4 = 7 apples. Josh and Mia made 4 + 3 = 7 cookies. The order in which you add the groups doesn't matter. You get the same answer.

Now You Try

How many more cookies would Josh and Mia have to make to be able to give 1 cookie each to 10 people? Use buttons to help figure this out. ➤

13

I'm at the beach with Luis.
We are going to make sandcastles
and we want to decorate them with
seashells. Luis has 7 seashells and
I have 3 more seashells than him.
Can we use addition to find out how
many seashells I have collected?

That's a good reason to use addition.
You know Luis has 7 seashells.
Now count 3 more. What's the answer?

14

Josh had 3 more seashells than Luis. We know how many seashells Luis had, so we can figure out how many Josh had.

When I add 7 and 3, I get 10.

$$7 + 3 = 10$$

Now I will count the seashells on my sandcastle and see if I get the same answer. I was right! I have 10 seashells, 3 more than Luis.

Now You Try

If Luis had 5 seashells and Josh had 3 more, how many seashells would Josh have? Draw the seashells on a piece of paper and count them.

I'm having a party and inviting 9 of my friends, so there'll be 10 people all together. I want to buy enough cans of soda for each person to have 1 can. There are only 6 cans on the shelf. How many more cans will I need?

Okay, Josh. Count up from 6 holding up one finger as you say each number. Then all you have to do is count how many fingers you are holding up.

Up to 10

Don't forget you can use your fingers to figure out problems that have numbers up to 10.

That's easy! If I count 7, 8, 9, 10, then I'm holding up 4 fingers. So I need to ask the store manager for 4 more cans of soda.

$$6 + 4 = 10$$

The store manager had 4 more cans in his storeroom, so there's enough for everyone.

Now You Try

If Josh found 4 cans of soda on the shelf, how many more would he need to have 10 all together?

He would need 6 more

We want to start a school band. We have 2 drums, 3 trumpets, and 4 flutes. Each band member will play 1 instrument. How many people can join the band?

This time we have 3 sets to add, but we still use addition. We start by adding 2 of the sets. Then we add their sum to the other set.

Start by adding the drums and the trumpets.
2 + 3 = 5. Then add this number to the
4 flutes, and that will be the number of
people in the band.

That's easy, Digit. I know that 5 and 4
make 9, so there are 9 people in the
school band.

$$5 + 4 = 9$$

Now all we have to do
is decide what tune to play!

Adding More than 2 Sets

Adding more
than 2 sets is as
easy as adding 2.
First add 2 sets,
then add the
answer to the
next set. Breaking
the problem
down means you
only have to add
2 things at a time.

Now You Try

If there were 3
trumpets, 5
drums, and
2 flutes,
how many
people would be
in the band?

It's Halloween. My dad is taking Mia, Luis, and me trick-or-treating. We've been given 5 lollipops, 6 jellybeans, and 9 candy bars. How much candy do we have all together?

Let's start by adding the 5 lollipops and the 6 jellybeans. Use counters or buttons to help you and count each one. 6 plus 5 makes 11.

10 people

20

Okay, Digit. I can do the rest. I have 9 candy bars, so I need to add 9 to 11. That makes 20.

$$5 + 6 + 9 = 20$$

That's a lot of candy! We'd better not eat it all at once.

Checking Your Answer

Whichever way you add the 3 sets, you should get the same answer. You can check the answer by starting with 2 different sets. $(5 + 6) + 9$ is the same as $(6 + 9) + 5$ and $(5 + 9) + 6$.

Now You Try

If the children had 2 jellybeans, 4 lollipops, and 8 candy bars, how many candies would they have all together? Check your answer by adding the 3 sets another way.

21

It's Holly's birthday, and I want to give her a bunch of flowers. I've been growing some flowers in my garden. There are 3 rows of flowers. In one row there are 6 red flowers. In the next row there are 6 white ones, and in the last row there are 6 yellow ones. I'm going to pick 2 red flowers, 3 white flowers, and 4 yellow flowers to give to Holly. How many flowers will be left in the garden?

Okay, first pick the flowers, then count how many flowers are left in each row. Do you know what you should do next?

14 candies

22

Sure, Digit. There are 4 red flowers, 3 white flowers, and 2 yellow flowers still growing in their rows. So if I want to know how many flowers are left in the garden, I can add these together.

$$4 + 3 + 2 = 9$$

There are 9 flowers left in the garden. I hope Holly likes her birthday present.

Opposite

When you add, you join 2 or more sets together to make a bigger set. There's another type of math called subtraction. When you subtract, you take small sets away from a big set. Subtraction is the opposite of addition.

Now You Try

Can you write an equation for the number of flowers in Holly's bunch?

23

Brad and I are wearing our special number T-shirts. My T-shirt has the number 7 on it, and Brad's T-shirt has the number 8. The balloons have numbers painted on them, but the strings are all tangled. How do we know which balloons we should be holding?

You can find out which balloons you should be holding by solving some equations. The numbers on the balloons add up to 7 or 8. You should be holding the balloons that add up to 7 and Brad should be holding the balloons that add up to 8.

$$2 + 3 + 4 = 9$$

Addition Facts

Addition facts are pairs of numbers that can be added together to give another number. Brad and Josh found the addition facts for 7 and 8, but all numbers have them.

I can see that one of the balloons has 0 on it. It's important to learn that adding 0 to a number doesn't change the number.

My T-shirt has the number 7, so my balloons say "2 + 5," "3 + 4," and "6 + 1." They all add up to 7. Brad's T-shirt has the number 8, so he should hold the balloons that say "3 + 5," "6 + 2," and "8 + 0."

We'd better not get the strings tangled again.

Now You Try

Two of the addition facts for 4 are "0 + 4" and "4 + 0." Can you find the other addition facts for 4?

I've drawn the numbers from 0 to 20 on the sidewalk. Can I use this to find the number that must be added to 7 to make the sum of 15?

Yes you can, Josh. You've drawn a number line. This is very useful for finding the answer to lots of addition questions. Start at the 0 and hop along the numbers, counting as you go. From 0 to 7, there are 7 hops.

(3 + 1), (2 + 2), (1 + 3)

From 7, hop again until you reach 15.
How many hops is that?

*I counted 8 hops from 7 to 15, so 8 is the
other number in this addition fact for 15.
I can write this as an equation*

$$7 + 8 = 15$$

Number Line

You can draw your
own number line.
A number line goes
up to any number
you choose. Don't
forget to start at 0!

Now You Try

If you started
at 0 and
hopped 8
spaces, how
many more
spaces would
you need to hop
to get to 15?

When I deliver newspapers to the houses on my block, Mia and her dog Popcorn sometimes help me. All the houses on one side of the road have odd numbers. The houses on the other side have even numbers. What happens if I add odd numbers together? What happens if I add even numbers?

Start with the odd numbers. 1 + 3 = 4 and 3 + 5 = 8. What happens when you add a pair of odd numbers?

7 spaces. 8 + 7 = 15 and 7 + 8 = 15

28

Adding 2 odd numbers gives an even number. Now I'll try the even numbers.

$$2 + 2 = 4 \text{ , and } 4 + 6 = 10$$

So adding 2 even numbers gives an even number, too. I'll remember that!

Now You Try

What happens if you add an odd number and an even number? Try these pairs 3 + 2, 3 + 4, 5 + 4, 3 + 6. Are the answers even or odd?

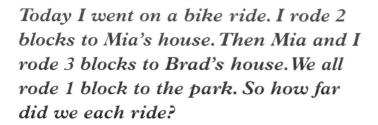

Today I went on a bike ride. I rode 2 blocks to Mia's house. Then Mia and I rode 3 blocks to Brad's house. We all rode 1 block to the park. So how far did we each ride?

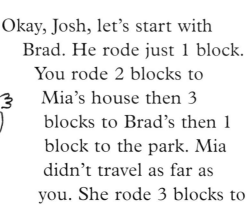

Okay, Josh, let's start with Brad. He rode just 1 block. You rode 2 blocks to Mia's house then 3 blocks to Brad's then 1 block to the park. Mia didn't travel as far as you. She rode 3 blocks to Brad's house then 1 more block. Can you figure it out?

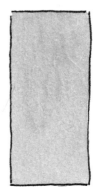

The answers are all odd

Sure, Digit. I just need to solve some addition equations

$$2 + 3 + 1 = 6, \text{ and } 3 + 1 = 4$$

So I rode 6 blocks and Mia rode 4 blocks. No wonder I'm so tired.

Adding Distances

Josh used addition to figure out how far he and his friends rode. Addition isn't just about things like cookies and candies and mice. We can also add things that we cannot see or touch, like distance and time.

Now I know that addition is all around me. It helps me count runaway mice, figure out how many cookies I've made, and find out how many people can join the school band. I know that addition is about joining groups of things together to make a bigger group. I can check my answer to make sure I've got it right. And I can see how far I've biked with my friends!

Here are some useful addition words

Addition facts: Two numbers that add together to give another number are called addition facts.

Equation: An equation is like a sentence in math. It uses numbers and special signs instead of words.

Even numbers: Even numbers end in 2, 4, 6, 8, or 0.

Odd numbers: Odd numbers end in 1, 3, 5, 7, or 9.

Set: A set is a group of things. Addition is about joining 2, 3, or more sets together.

Subtraction: This is a type of math that lets you take smaller sets away from a big set.

Sum: The answer to an addition question is called a sum.